Fluid Power
Educational
Series

Hydraulic Rotary Actuators

(In the English Units)

Joji Parambath

Hydraulic Rotary Actuators
(In the English Units)

Copyright © 2026 Joji Parambath

All rights reserved

ISBN: 9798653847790

https://fluidsys.org

First Edition 2020
Revised Edition 2026

Disclaimer of Liability
The contents of this book have been checked for accuracy. Since deviations cannot be precluded entirely, we cannot guarantee full agreement. Only qualified personnel should be allowed to install and work on pneumatic and hydraulic equipment. Qualified persons are defined as persons who are authorized to commission, ground, and tag circuits, equipment, and systems following established safety practices and standards.

Table of Contents

Chapter	Description	Page No
	Preface	v
1	Basic Hydraulic Motor Working	1
2	Differences between Pumps and Motors	3
3	Terms and Definitions – Hydraulic Motors	4
4	Constructional Features of Hydraulic Motors	15
5	Side Loads on Hydraulic Motors	22
6	Classification of Rotary Actuators	24
7	Semi-rotary Hydraulic Actuators	26
8	Hydraulic Motors	29
9	Comparison of Hydraulic Motors	43
10	Performance Characteristics of Hydraulic Motors	45
11	Applications of Hydraulic Motors	48
12	Maintenance of Hydraulic Motors	49
13	Objective Type Questions	51
14	Review Questions	53
15	Numerical Problems	56
Appendix 1	(a) Specifications – Gerotor Motor (b) Specifications – Geroler Motor	57
Appendix 2	Typical Specifications - Bent-axis Axial Piston Motor	66
16	References	67

Preface

Rotary actuators are the muscle behind the rotary motions in industrial and mobile hydraulic systems. Information and data on hydraulic motors are scattered across many existing textbooks and manufacturer documents. A modest attempt is made here to present the information and data in a single book, in simple language and a structured way.

The book presents the fundamentals and other essential technical information on hydraulic motors. The topics are logically arranged in a simple-to-complex progression. The book uses the English system of units.

Many other fluid power topics are covered in other textbooks in the same author's fluid power educational series. A list of all the books is given at the end of the book. Also, please see the details at https://jojibooks.com.

Enjoy reading the book.
Your feedback is most welcome.

JOJI Parambath

Chapter 1 | Basic Hydraulic Motor Working

Hydraulic rotary actuators are positive-displacement devices that convert hydraulic energy to rotary mechanical energy. A rotary actuator, used in a hydraulic system, converts system pressure and flow into controllable rotary force (torque), rotary motion, or both. Here, the fluid pressure is converted into torque, and the flow rate into rotary speed. The torque and/or motion of the rotary actuator can be used to achieve rotary motion in industrial machinery.

Figure 1.1 | A graphic representation of a hydraulic motor*
Courtesy: Penton Business Media, Inc., U. S. A.
*Note: "The graphic is the copyrighted property of and is reprinted with the permission of Penton Business Media, Inc."

Hydraulic rotary actuators can be classified into two types. They are: (1) Semi-rotary actuators and (2) Motors. A semi-rotary actuator is capable of producing only limited rotation and can twist objects along a partial arc. On the other hand, a hydraulic motor can produce continuous rotation and impart continuous rotary motion to the connected load. Figure 1.1 shows the cut-section view of a hydraulic motor.

Basic Motor Operation

A hydraulic motor consists mainly of moving elements, such as gears, vanes, or pistons, connected to the motor's output shaft and enclosed in a single housing.

Consider that the motor is employed in a hydraulic system. The shaft rotates when the pressurized system fluid is applied to the motor's rotating parts. In this way, the motor can convert the applied pressure into rotary mechanical force and, consequently, drive the load attached to it. The fluid then returns to the system reservoir after passing through the motor. A drain connection is provided in the motor to return leaked fluid to the reservoir.

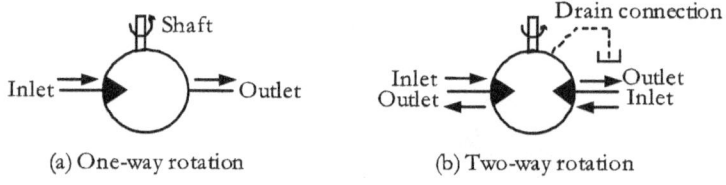

(a) One-way rotation   (b) Two-way rotation

Figure 1.2 | Schematic diagrams of hydraulic motors

Figures 1.2(a) and (b) show the illustrative diagrams of unidirectional and bidirectional hydraulic motors, respectively. The unidirectional hydraulic motor provides rotation in only one direction, whereas the bidirectional hydraulic motor can rotate in both clockwise and anticlockwise directions. The direction of rotation of the motor's shaft can easily be reversed by changing the direction of the fluid flow through the motor ports.

This book describes various aspects of semi-rotary actuators and hydraulic motors, including their construction, classification, and operation.

Chapter 2 | Differences between Pumps and Motors

Understandably, hydraulic motors are very similar to hydraulic pumps in design and construction. However, there are many differences between the two. The main difference between a pump and a hydraulic motor is that, in a pump, the moving parts, connected to a hydraulic system, push fluid through the system to create flow and pressure. In contrast, in a motor, the pressurized fluid pushes its moving elements to produce rotary mechanical motion and force.

Another difference is that the pump is always coupled to its prime mover, whereas the motor is always coupled to a load. In addition, some design modifications are required for hydraulic motors due to their specific application requirements.

For example, a hydraulic motor has to overcome high starting torque at low speeds and the effects of side loading. The symbolic representations of a hydraulic pump and hydraulic motor are given in Figure 2.1.

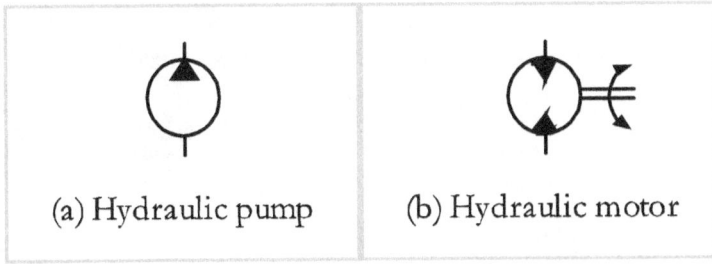

(a) Hydraulic pump   (b) Hydraulic motor

Figure 2.1 | Symbolic representations of a hydraulic pump and hydraulic motor

Chapter 3 | Terms and Definitions –
Hydraulic Motors

Some critical factors relevant to the operation and applications of hydraulic motors include operating pressure, displacement, flow rate, input power, output power, torque, and efficiency. The following sections describe these terms.

Operating Pressure (P): It is the pressure in a hydraulic system that overcomes all resistances, including both useful work and losses. The rated pressure of a hydraulic motor is the maximum pressure that the manufacturer recommends for the motor.

Motor Displacement (V_D): It refers to the volume of the system fluid required for turning the output shaft of a motor through one revolution. Some of the units of motor displacement are m^3/rev, cc/rev, or in^3/rev.

Theoretical Flow Rate (Q_T): It is the quantity of the system fluid that must flow through a motor per unit of time, provided there is no leakage in the system. In the English system of units, the flow rate is measured in cubic inches per minute (in^3/min) or gpm. The mathematical equations for the theoretical flow rate (Q_T) of the hydraulic motor in the English system of units are as follows:

$$Q_T(gpm) = \frac{V_D(in^3/rev) \times N(rpm)}{231}$$

Slippage in Hydraulic Motors: It is the internal leakage of the system fluid that flows through unintended paths within a motor, without performing any useful work. As slippage in the hydraulic motor increases, more of the available flow intended for useful work is lost, reducing motor power.

However, all hydraulic motors are susceptible to some amount of slippage. Slippage increases as system pressure rises.

Other reasons for the increase in slippage are wear-induced enlargement of the clearances between the motor's internal parts and temperature-induced thinning of the system fluid.

Speed: It is directly related to the theoretical flow rate of a motor and inversely related to the displacement of the motor. It can be expressed in the English system of units by the following equation:

$$\text{Speed, N (rpm)} = \frac{\text{Theoretical motor flow, } Q_T \text{ (gpm) x } 231}{\text{Motor displacement, } V_D \text{ (in}^3/\text{rev)}}$$

Therefore, it can be observed that, for a given flow rate, increasing the motor displacement decreases the motor speed and vice versa. Remember that the intended application determines the hydraulic motor's operating speed.

Maximum motor speed is the speed of a hydraulic motor, at a particular inlet pressure, that it can sustain for a limited period without damage to the motor.

Minimum motor speed is the slowest, continuous, rotational speed obtainable from the output shaft of a hydraulic motor.

Input Power (P_{in}): Figure 3.1 shows the block diagram of a hydraulic motor, with the power relationships at its input and output sides.

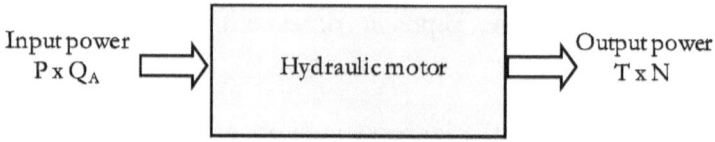

Input power
$P \times Q_A$

Hydraulic motor

Output power
$T \times N$

Figure 3.1 | A block diagram showing the power relationships in the hydraulic motor

The mathematical equation for the input power of the hydraulic motor, in the English system of units, is as follows:

$$\text{Input Horse Power (hp)} = \frac{P(\text{psi}) \times Q_A(\text{gpm})}{1714}$$

Theoretical Torque (T_T): Theoretical torque of a hydraulic motor is a function of the motor's displacement and the system pressure. The following section presents the mathematical equation for the motor's theoretical torque in English units. The theoretical figures represent the torque available at the motor shaft, assuming no mechanical losses.

$$\text{Theoretical Torque, } T_T(\text{in.lb}) = \frac{V_D(\text{in}^3/\text{rev}) \times P(\text{psi})}{2\Pi}$$

Breakaway (Starting) Torque: It is the rotary force required to turn a stationary load connected to the motor.

More torque is required to turn the stationary load than that required to keep it moving. This is because the motor must initially overcome the inertia of the load. Therefore, the motor needs a breakaway (starting) torque large enough to overcome the load torque.

Running Torque: It is the torque required to rotate a load connected to the motor. It is determined by the motor's internal volumetric displacement and the pressure drop across it. Remember, the pressure relief valve setting limits the hydraulic motor's running torque.

Stalling Torque: The minimum load torque required to stop a running hydraulic motor is its stalling torque. In this condition, the maximum hydraulic torque, limited by the pressure relief valve setting, is insufficient to overcome the load torque.

Torque Ripple: It is the difference between the minimum torque and maximum torque delivered by the motor at a given pressure during its one cycle of rotation.

Example 3.1 | A skid steer broom used to clean a construction site has a hydraulic motor with a displacement of 3.2 in3/rev and an operating pressure of 3000 psi. What is the maximum theoretical torque the motor is capable of producing?

Solution

Volumetric displacement, V_D = 3.2 in^3/rev
Pressure, P = 3000 psi

Theoretical Torque, T_T
= V_D (in^3/rev) x P (psi)/(2$\prod$) in.lb
= 3.2 x 3000 / (2π) = 1527.89 in.lb]

Actual Torque (T$_A$): It is the torque that a motor develops to drive the attached load alone. It is equal to the theoretical torque minus the torque losses on account of any friction in the motor.

Output Power (P$_{out}$): The mathematical equation for the output power of a hydraulic motor, in the English system of units, is as follows:

$$\text{Output Horse Power (hp)} = \frac{T_A(\text{lb.in}) \times N \,(\text{rpm})}{63025}$$

Motor Efficiency: The efficiency of a hydraulic motor is the ratio of its output power to its input power. An ideal hydraulic motor has no leakage or frictional losses and is 100% efficient.

In practice, however, the motor has leaks and frictional losses. Accordingly, two basic types of efficiencies are identified for the motor. They are:

- Volumetric efficiency
- Mechanical efficiency.

Overall efficiency can then be derived from these two types. The following sections briefly explain these terms.

Volumetric Efficiency (η_v) of the hydraulic motor is the ratio of the theoretical flow rate responsible for developing the actual motor speed to the total flow rate consumed by the motor, including the leakage in the motor.

Remember, the motor draws more flow than it should due to leakage.

The mathematical equation for the volumetric efficiency of the motor is as follows:

$$\text{Volumetric efficiency, } (\eta_v) = \frac{\text{Theoretical flow rate } (Q_T)}{\text{Actual flow rate } (Q_A)}$$

Remember, the speed of a hydraulic motor depends entirely on the flow through it and is independent of the pressure drop across it.

Mechanical efficiency (η_m) of the hydraulic motor is the ratio of the actual torque delivered by the motor to the theoretical torque of the motor. The hydraulic motor produces less torque than it should due to frictional losses. The mathematical equation for the mechanical efficiency of the hydraulic motor is as follows:

$$\text{Mechanical efficiency, } (\eta_m) = \frac{\text{Actual torque, } (T_A)}{\text{Theoretical torque } (T_T)}$$

Overall Efficiency (η_o) of the hydraulic motor is the ratio of the 'brake' power delivered by the motor to the hydraulic power delivered to the motor. Both volumetric and mechanical efficiencies reduce the motor's overall performance. Therefore, the overall efficiency of the motor is also the product of its volumetric efficiency and its mechanical efficiency and is expressed mathematically as:

$$\text{Overall efficiency, } (\eta_o) = \frac{\text{Brake power delivered by the motor}}{\text{Hydraulic power delivered to the motor}}$$

$$= \eta_v \text{ x } \eta_m$$

The relationship summary is presented in the next section.

Summary of Relations for Hydraulic Motors

Figure 3.2 summarizes the essential relations of hydraulic motors in English units.

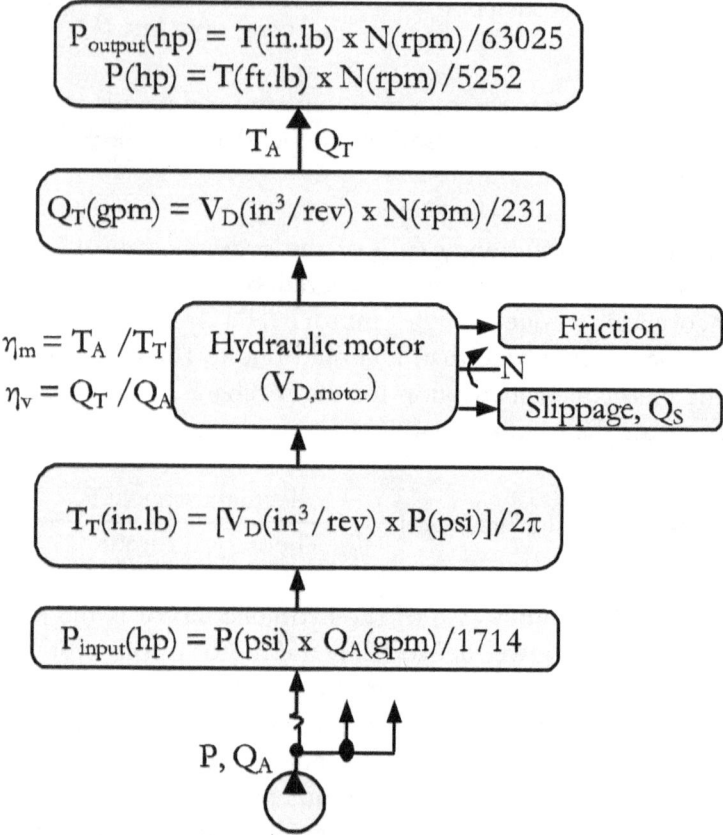

$$P_{output}(hp) = T(in.lb) \times N(rpm)/63025$$
$$P(hp) = T(ft.lb) \times N(rpm)/5252$$

$$T_A \quad Q_T$$

$$Q_T(gpm) = V_D(in^3/rev) \times N(rpm)/231$$

$$\eta_m = T_A/T_T$$
$$\eta_v = Q_T/Q_A$$

Hydraulic motor $(V_{D,motor})$

Friction

N

Slippage, Q_S

$$T_T(in.lb) = [V_D(in^3/rev) \times P(psi)]/2\pi$$

$$P_{input}(hp) = P(psi) \times Q_A(gpm)/1714$$

$$P, Q_A$$

Figure 3.2 | Summary of relations for a hydraulic motor

The mathematical relationship in a hydraulic motor describes the transformation of hydraulic power, determined by flow rate and pressure, into mechanical power in the form of torque and rotational speed.

Example 3.2 | A hydraulic gear motor consumes 11.7 gpm while running at a speed of 500 rpm. Assume the motor's volumetric efficiency to be 90%. What is the volumetric displacement of the motor?

Solution

Actual flow rate, Q_A = 11.7 gpm
Motor speed, N = 500 rpm
Motor speed, n = 8.33 rps
Volumetric efficiency, η_v=0.9

Theoretical flow rate, Q_T =11.7 x 0.9 = 10.53 gpm

Volumetric displacement, V_D =Q_T (gpm) x 231 /N (rpm)
= 10.53 x 231 / 500
= 4.86 in3/rev]

Example 3.3 | What is the actual torque supplied by a hydraulic motor of 3.05 in³/rev at 3626 psi? Assume the mechanical efficiency of the motor as 90%.

Solution

V_D = 3.05 in³/rev
Pressure, P = 3626 psi
Mechanical efficiency, η_m = 90%

Theoretical torque, T_T = V_D (in³/rev) x P (psi)/ (2$\prod$) in.lb
= 3.05 x 3626/ (2$\prod$)
= 1760 in.lb

Actual torque, T_A = η_m x T_T
=0.9x1760 in.lb = 1584 in.lb

Example 3.4 | A hydraulic motor rotates at a speed of 450 rpm with a nominal displacement of 0.244 in³/rev. The pressure differential across the hydraulic motor is 1088 psi. The overall efficiency is 80%, and the volumetric efficiency is 90%. Calculate the following: (1) Theoretical flow rate, (2) Actual flow rate, (3) Power input, (4) Shaft power, and (5) Shaft torque.

Solution

Speed, N = 450 rpm

Displacement, V_D = 0.244 in³/rev

ΔP = 1087.785 psi

η_v = 90%

η_o = 80%

Speed, n = 450/60 rps = 7.5 rps

Theoretical flow rate, Q_T = V_D (in³/rev) x N(rpm)/231

= (0.244 x 450)/231

= 0.475 gpm

Actual flow rate, Q_A = Q_T/ η_v

= 0.475/0.9 = 0.528 gpm

Power input P_{in} = ΔP (psi) x Q_A (gpm)/1714

= 1088 x 0.478/1714

= 0.3 hp

Shaft power, P_{out} = P_{in} x η_o

= 0.3 x 0.8 = 0.243 hp

Shaft torque, T = P_{out} (watt) /2π n

= (P_{out} (hp) x63025)/N(rpm)

= (0.243x63025)/450 = 34 in.lb

Example 3.5 | A hydraulic motor operating at a pressure of 1450 psi produces a displacement of 10 in³ and a speed of 2000 rpm. If the actual flow consumed by the motor is 95 gpm and the motor's actual torque is 2213 in.lb, find its output power, volumetric efficiency, mechanical efficiency, and overall efficiency.

Solution
$P = 1450$ psi
$V_D = 10$ in^3
$N = 2000$ rpm $= 210$ rad/s
$Q_A = 95$ gpm
$T_A = 2213$ in.lb

Output power $= T_A$ (in.lb)xN(rpm)/63025
$= 2213$x$2000/63025 = 70.23$ hp

Theoretical flow rate, $(Q_T) = V_D$ (m^3/rev) x n (rps)
$= V_D$(in^3/rev)xN (rpm)/231
$= 10$ x $2000/231$
$= 86.58$ gpm

Theoretical torque, $(T_T) = V_D$ (in^3/rev) x P (psi)/ $(2\prod)$
$= 10$ x $1450/$ $(2\prod)$
$= 2308$ in.lb

Mechanical efficiency, $(\eta_m) = T_A/ T_T$
$= 2213/2308 = 0.96$

Volumetric efficiency, $(\eta_v) = Q_T/ Q_A$
$= 86.58/95 = 0.91$

Overall efficiency, $(\eta_v) = \eta_v$ x $\eta_m = 0.91$ x $0.96 = 0.87$

Example 3.6 | Calculate the flow rate (size) requirement of a hydraulic motor to be used in a hydraulic system for driving a load of 5 hp at 3000 rpm. The system pressure relief valve is set to 3000 psi. Assume a return line pressure of 100 psi, a mechanical efficiency of 88%, and a volumetric efficiency of 93% for the motor.

Solution
Output power, P_{out} = 5 hp
Motor speed, N = 3000 rpm
Motor speed, n = N/60 = 3000/60 = 50 rps
System pressure, P_{system} = 3000 psi
Return line pressure, P_{return} = 100 psi
Pressure differential, ΔP = 3000-100 = 2900 psi

Volumetric efficiency, η_v = 0.93
Mechanical efficiency, η_m = 0.88
Overall efficiency, η_o = η_v x η_m = 0.93 x 0.88 = 0.82

Input power, P_{in} = P_{out} / η_o
=5 / 0.82 = 6.1 hp

Theoretical torque, T_T = (P_{in} (hp) x 63000)/ N
= 6.1 x 63000/3000 = 128 in.lb

Motor displacement, V_D = T_T x $2\prod/$ (ΔP)
= 128 x $2\prod/$ (2900) = 0.277 in^3/rev

Theoretical flow rate, Q_T = V_D (m^3/rev) x n (rps)
= V_D (in^3/rev) x N(rpm) /(231)
= 0.277 x 3000/231 = 3.597 gpm]

Actual flow rate, Q_A = Q_T/ η_v =3.597/0.93 =3.87 gpm

Chapter 4 | Constructional Features of Hydraulic Motors

The general constructional features of hydraulic motors are similar to those of hydraulic pumps. However, some design modifications are required because of their unique application requirements. For example, a hydraulic motor has to overcome high starting torque at low speeds and the effects of side loading.

The nature of applications usually dictates the construction materials of hydraulic motors. These construction materials include cast iron, ductile iron, bronze, cast steel, and stainless steel. The body and shaft of a hydraulic motor can be nickel-plated for highly corrosive environments or food/sanitary applications.

A speed sensor can be integrated into a hydraulic motor to obtain speed and rotational direction data. The shafts, ports, and mounts of hydraulic motors conform to ISO or SAE standards. The unique features of hydraulic motors include shaft seals, seal guards, speed sensors, case drain ports, case drain internal check valves, integrated flushing (shuttle) valves, and integrated brake valves.

High-pressure Shaft Seal
A high-pressure shaft seal (rotary seal) is a critical element of a hydraulic motor. It is placed in the matching groove on the motor part. The housing to which the sealing is fitted and the shaft that rotates within it constitute a sealing system. The primary purpose of the sealing system is to prevent the system fluid from leaking through any clearances between the motor's mating parts.

A high-pressure shaft seal is designed with a specialized geometry for applications operating in extreme conditions. The seal's geometry increases the clamping force of the sealing lip against the output shaft, preventing seal leakage under high-pressure conditions. An ideal sealing system requires a perfectly concentric shaft. However, this requirement is difficult to achieve, and some eccentricity is inevitable. A seal compensates for this eccentricity.

When case pressure is developed, the seal lip clamps more tightly against the output shaft, resulting in a reliable sealing system that can handle high-pressure spikes without failing. The seal can withstand case pressures of up to 290 psi at 1500 rpm, depending on the seal material.

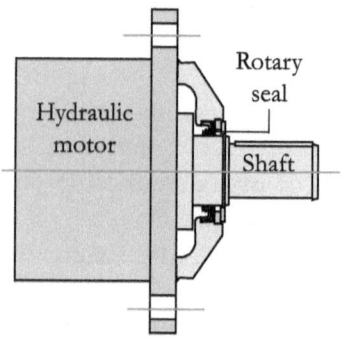

Figure 4.1 | Shaft seal

Shaft seals are available in various sizes, profiles, and materials. Seal materials include NBR and FPM. The nominal lifespan of the shaft seal can be anticipated at a specified speed and the associated maximum case pressure. However, the seal's longevity may be diminished under adverse operating conditions, such as elevated temperatures, reduced oil viscosity, and contaminated oil.

Seal Guard

A seal guard is a metal shield that protects the internal wiper seal of a hydraulic motor. It is installed on the output shaft and moves along with it. The seal guard is usually recessed into a groove in the bearing housing face. It protects the shaft seal from physical damage caused by external contamination. This component is especially used in hydraulic motors used in various types of equipment, such as mining machinery, harvesting machines, industrial sweepers, and lawn equipment.

Speed Sensor

A speed sensor provides a digital signal over a wide range of speeds and temperatures. Its ON/OFF signal can easily be interfaced with the associated control system.

Case Drain Connections

A hydraulic motor, especially the piston type, is usually designed with a case drain with ports to drain the leakage fluid to the associated system reservoir. A case drain line must be correctly sized and directly connected from the drain port to the reservoir, with no restrictions.

The line must be piped, siphoning must be prevented, and the termination must be below the minimum fluid level in the associated reservoir.

A hydraulic motor may also be designed to allow a small amount of internal leakage to lubricate and cool its internal parts.

The leakage fluid in a motor's case experiences pressure. The higher the case pressure, the shorter the shaft seal's life. The case pressure should be limited based on shaft speed and seal type.

For example, the maximum recommended case pressure is typically 290 psi at a shaft speed of 1500 rpm and an NBR seal, whereas it is 5.8 psi at a shaft speed of 6000 rpm and an FPM seal. Figure 4.2 shows the configurations of a hydraulic motor's drain ports when installed horizontally and vertically.

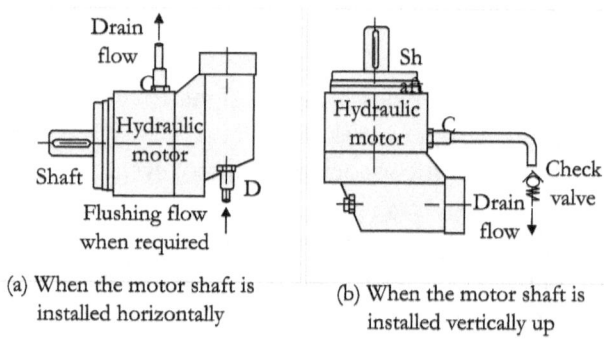

(a) When the motor shaft is installed horizontally

(b) When the motor shaft is installed vertically up

Figure 4.2 | Configurations of drain ports of hydraulic motors

Figure 4.1(a) shows how drain ports C and D are configured when the motor shaft is horizontal. The drain port C at the top must be used for draining. Ideally, this drain port should be directly connected to the tank. If flushing is needed, the flow should be directed from drain ports D to C.

Figure 4.1(b) shows the typical configuration of drain ports when the motor is installed with the shaft up. To ensure enough fluid in the case, a spring-loaded check valve should be installed in the drain line.

The case of a hydraulic motor should be filled with fluid before it starts up. The internal leakage, especially at low operating pressures, is insufficient to provide lubrication at start-up.

Case Drains for Gear Motor Designs

External case drains are essential for piston-type hydraulic motors. However, gear motors, like external, internal, and gerotor/geroler motors, usually have small-volume cases. Therefore, many gear motors lack case drains. However, they have small internal leakage, which usually accumulates in a small cavity just behind their shaft seal.

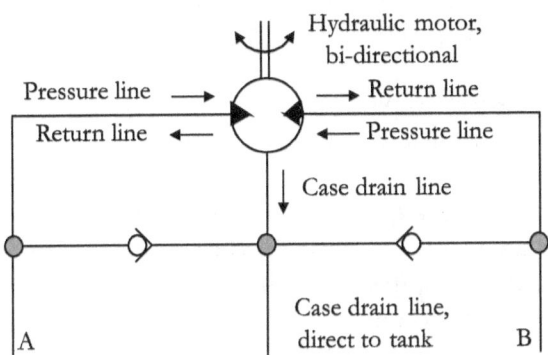

Figure 4.3 | Case drain for a gear motor

This leakage can be reduced and drained by installing a high-pressure shaft seal and two in-built check valves, as shown in Figure 4.3. Internal leakage can be relieved either directly or through the return line into the tank.

If the case drain is not directly connected to the reservoir, the case pressure can be assumed to equal the return-line pressure. Therefore, the motor's case pressure can be reduced.

The external case drain can be connected when the return-line pressure exceeds the recommended case drain pressure. This helps extend the motor seal life, improve cooling by drawing the heat away, and reduce contamination.

Integrated Brake Valve

A hydraulic motor used in a vehicle's hydrostatic transmission (HST) system tends to operate faster than the available system flow when the vehicle runs steeply downhill. This faster operation may lead to undesirable cavitation in the motor.

A brake valve can be integrated to throttle the motor's return flow and provide braking.

Anti-cavitation Valve

A hydraulic motor is susceptible to cavitation when the inlet pressure is insufficient. An anti-cavitation valve, as shown in Figure 4.4, can reduce the effects of cavitation. It is merely a check valve connected between the pressure and return ports. It opens to ensure flow to the motor if the inlet pressure becomes too low. Sufficient back pressure on the return line is essential. The motor must have a defined rotation direction when using the anti-cavitation valve.

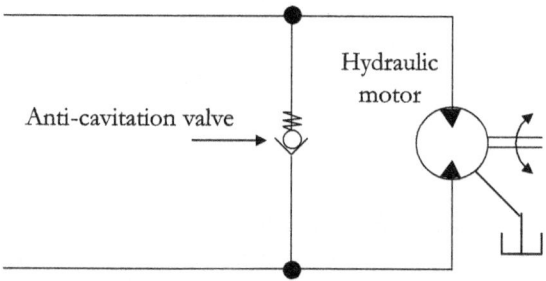

Figure 4.4 | Anti-cavitation valve connected to a motor

Integrated Flushing Valve

A hydraulic motor used in a hydrostatic transmission (HST) system operating at high speeds and power levels may have a separate or integrated flushing valve to provide additional cooling flow to its rotating parts. See Figure 4.5.

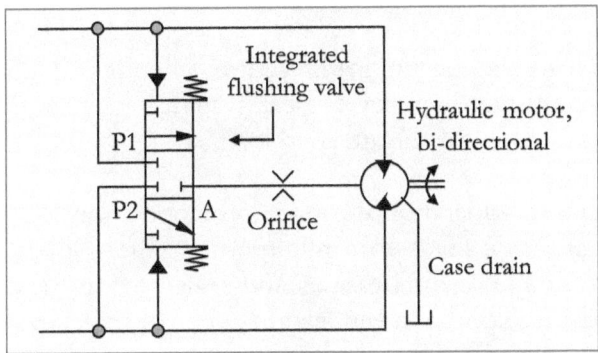

Figure 4.5 | A hydraulic motor with an integrated flushing valve

Case flushing of a hydraulic motor is required for its continuous high-speed operation to meet the viscosity and temperature limitations. Manufacturers publish data showing operating speeds above which flushing is usually required, as well as recommended flow through the case.

Integrated Parking Brakes

Hydraulic motors can be provided with integrated parking brakes. A spring-applied, pressure-release type parking brake holds a motor load by the spring when hydraulic pressure is absent. The brake can be released by hydraulic pressure. The brake pressure cavity is the motor case. As a result, maximum release pressure is limited by the case-pressure capability.

Chapter 5 | Side Loads on Hydraulic Motors

The performance of hydraulic motors is affected by thrust and side loads. A thrust load occurs in a hydraulic motor when a compressive load acts along the longitudinal axis of the motor shaft.

A side load also occurs in the motor when it is coupled to the load through a pulley or gear system, or when its shaft bears the weight of the attached load.

A crucial consideration during motor operation is keeping thrust and side loads to a minimum, as these loads affect the service life of the bearings and seals used in the motor. This adverse condition can lead to eventual shaft breakage.

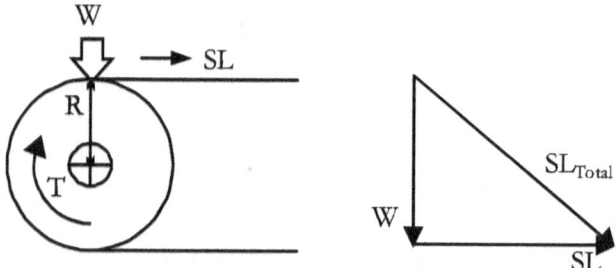

Figure 5.1 | A schematic diagram showing the side load components of a hydraulic motor

Figure 5.1 shows the sketch of the motor driving the load with a torque 'T' through a pulley of radius 'R'. Then, the side load (SL) on the motor due to the application of the load is given by:

$$SL = T/R$$

If the external load with a weight 'W' (Newton [lb]) acts on the motor shaft, then the total side load on the shaft is given by:

$$SL_{Total} = \sqrt{(W^2 + SL^2)}$$

The calculated side load on the motor at the given motor speed must be compared with the manufacturer's side load chart to determine whether the motor is within the permissible operating conditions. A motor of a larger physical size must be selected if the calculated side load goes beyond the permissible limit.

Side loads on the motor will fatigue the shaft and place a heavy load on the shaft bearing through leverage. Bearing and shaft life can be prolonged by keeping the distance between the load line and bearing centerline as short as possible.

Example 5.1 | The motor of a Brush mower produces a torque of 1062 in.lb to drive a pulley of 8 in in diameter at 375 rpm. If the load with a weight of 225 lb is also acting on the motor shaft, what is the total side load due to the weight and the side load on the motor?

Solution

Torque, T = 1062 in.lb
Diameter of pulley, D = 8 in
Weight acting on the shaft, W = 225 lb

Side load, SL = 1062 / (8/2) = 265.5 lb

Total side load, $SL_{Total} = \sqrt{(225^2 + 265.5^2)} = 348$ lb

Chapter 6 | Classification of Rotary Actuators

Hydraulic rotary actuators can be classified as semi-rotary actuators and motors based on their degree of rotation. Further, they are classified by several relevant parameters, such as the type of internal moving elements, the nature of displacement, and the torque/speed requirements of hydraulic rotary actuators. The following sections elaborate on these classifications.

Based on the type of their internal moving elements, hydraulic motors are classified as (1) Gear motors, (2) Vane motors, and (3) Piston motors.

According to the type of displacement, hydraulic motors are divided into fixed-displacement and variable-displacement types.

The fixed-displacement hydraulic motor displaces a fixed amount of system fluid with each revolution of the motor. Its displacement cannot be varied except by changing the flow rate of the system fluid.

A fixed-displacement motor typically produces a constant torque. Gear, vane, and piston motors can be designed for fixed-displacement operation. Remember, the speed of a motor is related to its displacement and the amount of fluid delivered to it.

The variable-displacement motor is constructed with an adjustment mechanism that can change its fluid displacement per revolution. With such an adjustment mechanism, the motor speed can be adjusted from zero to the maximum limit while keeping the pump delivery constant. The motor's torque can also be varied by varying its displacement. With the motor's input flow and operating pressure constant, varying its displacement can adjust the torque-to-speed ratio to suit the load requirements.

Only vane motors and piston motors can be designed and employed for the variable-displacement operation.

According to their torque-speed characteristics, hydraulic motors can be classified into two basic types.

One type is referred to as the high-speed, low-torque (HSLT) motor, and the other as the low-speed, high-torque (LSHT) motor.

The growing demand for hydraulic motors capable of driving high-inertia loads at low speeds led to an expansion of LSHT hydraulic motor technology.

A typical low-speed hydraulic motor can operate at speeds of 0.1 to 1000 rpm and has a starting torque of 75% to 90% of its maximum torque. LSHT motors are known for their reliability, high-power density, and modularity.

On the other hand, a typical high-speed hydraulic motor can operate at speeds from 1000 to 5000 rpm, but it typically lacks the high starting torque capability.

Typical high-speed motors are external gear motors, vane motors, in-line piston motors, and bent-axis piston motors.

Typical low-speed high-torque motors include gerotor motors, geroler motors, and radial piston motors.

Chapter 7 | Semi-rotary Hydraulic Actuators

A semi-rotary hydraulic actuator is a device that rotates its shaft using pressurized fluid through a fixed arc, usually less than 360°. The actuator's output torque varies directly with the fluid pressure applied to it. Semi-rotary hydraulic actuators can perform various operations, such as turning, tilting, transferring, indexing, mixing, clamping, feeding, and agitating. There are many designs of semi-rotary actuators. Figure 7.1 shows the symbol of a semi-rotary actuator, and the following sections explain the vane type, rack-and-pinion type, and helical gear-type semi-rotary actuators.

Figure 7.1 | Symbol of a semi-rotary actuator

Vane Type Semi-rotary Hydraulic Actuator

Figure 7.2 shows the schematic diagram of the vane type semi-rotary hydraulic actuator. It consists of one or more vanes enclosed in a cylindrical chamber with a block integral to its casing for separating its high-pressure side from its low-pressure side and the necessary inlet and outlet ports. The vanes are attached to the actuator shaft.

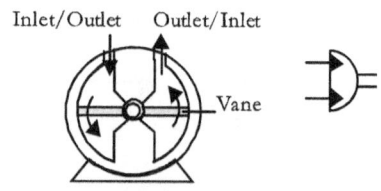

Figure 7.2 | A schematic diagram showing the cross-sectional view of a double-vane semi-rotary actuator

When the fluid enters at one port, the vane is pushed away in one direction, and the drive shaft is turned in that direction. When the fluid enters the second port, the vane is pushed away in the opposite direction, and consequently, the drive shaft is also turned in the opposite direction.

Rack-and-Pinion Type Semi-rotary Actuator

Figure 7.3 shows the schematic diagram of the rack-and-pinion type semi-rotary hydraulic actuator. It consists of a rack-and-pinion gear assembly and a cylinder part enclosed in shared housing. The pinion gear is attached to the actuator's driveshaft.

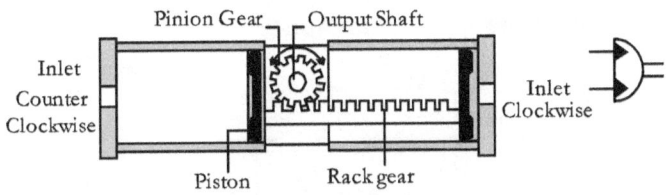

Figure 7.3 | A schematic diagram showing the cross-sectional view of a rack & pinion type semi-rotary actuator

The system fluid entering at one port of the actuator and exiting at the other port pushes the rack across the pinion gear. This action causes the shaft to rotate in one direction through a given arc.

Reversing the flow to the actuator ports reverses the shaft's direction of rotation. The torque produced in the actuator is a function of the piston area (A), the input pressure (P), and the radius of the pinion gear (R).

The rack-and-pinion type semi-rotary actuator can turn more than one revolution as its angle of rotation is in direct relation to the pinion gear size and the rack gear length. It may also be provided with cushions and stroke limiters for its smooth, adjustable stops.

Rack-and-pinion actuators are available in single- and dual-cylinder designs. The rotary motion of these actuators can be used in applications such as turning, tilting, indexing, mixing, and clamping.

Helical Gear Type Semi-rotary Actuator

Figure 7.4 shows the schematic of the helical-gear semi-rotary hydraulic actuator. In this type, a helical gear is machined in the drive shaft that meshes with a matching helical gear of the non-rotating piston.

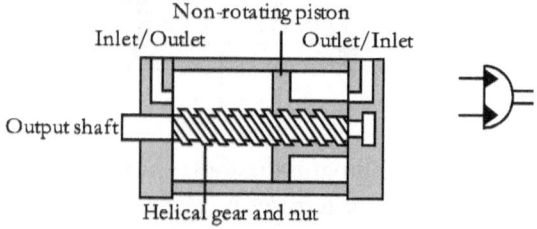

Figure 7.4 | A schematic diagram of a helical gear-type semi-rotary actuator

The fluid entering through one port imparts a twisting action to the output shaft in one direction. In the same way, the fluid entering through the other port imparts a twisting action to the output shaft in the opposite direction. The angle of rotation is in direct relation to the gear teeth angle and the non-rotating piston stroke.

This type of semi-rotary actuator can rotate its output shaft more than one revolution.

Chapter 8 | Hydraulic Motors

A hydraulic motor is a positive-displacement device that continuously turns its rotating elements and shaft using pressurized fluid in the associated system. The motor shaft can be connected to the load either directly or through clutches and gears. The motor can develop an output torque by allowing the system pressure to act on its moving parts.

The fluid entering through the inlet port pushes the moving parts. As a result, the shaft rotates and develops a torque. The torque developed by the motor depends on the system pressure, the area of the moving parts exposed to the pressurized fluid, and the distance between the moving elements and the shaft centerline. Internal motor leakage returns to the system reservoir through a case drain.

In general, the speed of a hydraulic motor is limited by its displacement and port size. The speed of a constant-displacement hydraulic motor remains constant for a constant input fluid flow. At the same time, the shaft torque of the motor remains constant as long as the differential pressure across the motor remains constant. Therefore, the motor's speed can be controlled by varying the flow rate, and the shaft torque can be varied by regulating the fluid pressure.

Many motors are designed for low-speed, high-torque (LSHT) operation, and some are designed for high-speed, low-torque (HSLT) operation. LSHT motors can transmit high torque despite their relatively small envelopes.

Hydraulic motors have many advantages. They run smoothly at low speeds. They can be stalled or used for rapid oscillatory motion without sustaining any damage. Moreover, they have high energy efficiency and operate at low noise levels.

Gear Motors

Gear motors are positive-displacement devices that produce torque by allowing the fluid pressure to act on their gears. By design, they are always fixed-displacement motors. The constructional features of the gear motors are similar to those of the gear pumps, but their operation is the opposite. That is, every gear pump pushes the system fluid to create pressure, whereas the pressurized fluid pushes every gear motor to develop torque. There are two types of gear motors with extensive applications across various machines. They are (1) external-gear motors and (2) internal-gear motors.

Gear motors have many advantages and some disadvantages. In general, they are the least expensive of all hydraulic motor types. They can tolerate contamination in their fluid media better than other hydraulic motor types. However, their efficiencies tend to be lower at lower speeds.

External-gear Motor

Figure 8.1 shows the schematic diagram of an external gear motor. It consists of a pair of matched gears enclosed in a housing, an inlet port, an outlet port, and a drive shaft. One gear unit is connected to the motor's output shaft, and the other one is an idler.

System fluid enters the housing through the inlet port. The fluid flows along the periphery of the housing, forcing the gears to rotate. Finally, it exits through the outlet port at a low pressure. The motor's rotary force is transmitted through its output shaft to perform useful work. The tight-fitting gears help control internal fluid leakage in the motor and increase its volumetric efficiency.

External gear motors are very compact and less expensive than piston-and-vane motors. They work best in high-speed operations and are suited for the bi-directional operation. The slippage losses in them are reasonably uniform for speeds above 500 rpm. Moreover, they can deliver relatively constant torques.

However, they are the noisiest and the least efficient of the three types of hydraulic motors.

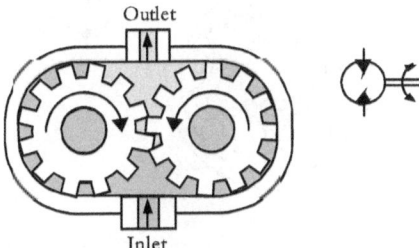

Figure 8.1 | A schematic diagram showing the cross-sectional view of an external gear motor

Gear motors are usually designed for high-speed, low-torque (HSLT) operation. They are suitable for applications with light starting loads but high running loads. They are very attractive as drives in many agricultural, construction, and mining equipment.

Gerotor/Geroler Motors

These types of motors are positive-displacement and categorized as internal-gear motors. They find extensive applications in mobile hydraulics.

Gerotor Motor

This motor arrangement, as shown in Figure 8.2, can be considered as a gear-within-a-gear type motor. It consists of two sets of interlocking gears – one set on the inner rotor and another set on an oblong rotor ring, with inlet and outlet ports, and an output shaft. The inner rotor has one tooth less than the rotor ring. The inner rotor is mounted on the motor's drive shaft and is eccentric to the rotor ring. Each tooth of the rotor gear is in contact with the internal surface of the rotor ring at all times. Fluid chambers are formed between the gear teeth and the housing.

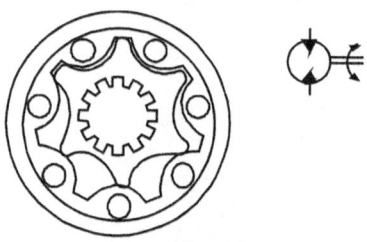

Figure 8.2 | Schematic diagrams showing the cross-sectional views of internal gear motors

A typical gerotor motor has a 7-tooth outer ring and a 6-tooth rotor, thus forming six fluid chambers. Consider that this motor is used in a hydraulic system. At any point in time, three consecutive fluid chambers at one end are pressurized, and the three consecutive chambers at the other end are connected to the system reservoir. The high-pressure fluid is delivered to the motor through the inlet port, where it flows through the fluid chambers. As the fluid passes through the fluid chambers, both gears rotate. The motor drive coupling transmits the rotor's motion to its output shaft. The fluid finally returns to the reservoir through the outlet port.

Gerotor motors can deliver considerable output power over a wide speed range. They are the most common type of low-speed, high-torque (LSHT) hydraulic motors used in industrial and mobile applications. They are used in agricultural and forestry equipment, construction machines, food processing machines, machine tools, lawnmowers, road rollers, excavators, and winches. Compact Gerotor motors are the natural choice for many applications, including plastic injection moulding, CNC tool changer drives, conveyor drives, clamping, and drilling.

Geroler Motor

A variant of the gerotor motor is the Geroler motor. Figure 8.3 shows the motor schematic.

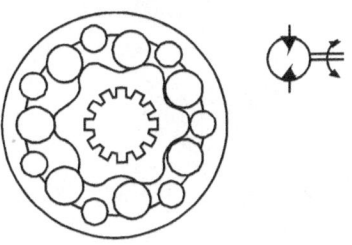

Figure 8.3 | Geroler motor

The operation of the Geroler motor is similar to that of the gerotor motor. However, the motor has its teeth fitted with rollers to reduce friction. This type of motor typically provides higher torques at lower speeds. The typical characteristics of Geroler motors include their compact, heavy-duty construction, smooth running even at low speeds, and reduced pressure spikes. These features give them a better efficiency even under high pressures and harsh operating conditions, and a long service life.

Geroler motors are well-suited for applications requiring quick start and stop cycles and rapid speed reversals. An application with a speed below 100 rpm should consider using a Geroler motor. Typical applications of this type of motor include agricultural and forestry equipment, construction machinery, material handling, lifting gear and winches, machine tools, and many other areas of industrial and mobile hydraulics.

Note: Appendix 1 gives the performance data for Geroler motors.

Vane Motor

A vane-type hydraulic motor is essentially a positive-displacement device. Figure 8.4 shows the schematic diagram of the vane motor.

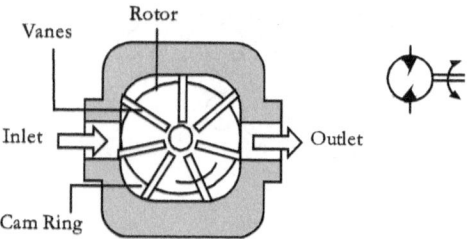

Figure 8.4 | A schematic diagram showing the cross-sectional view of a vane motor

The constructional features of a vane motor are very similar to those of a vane pump. However, the operation of the vane motor is just the reverse of that of the vane pump. That is, the vane pump produces hydraulic power in response to the rotary mechanical power at its drive shaft, whereas the vane motor produces rotary mechanical power at its drive shaft in response to the applied hydraulic power. The vane motor consists of a slotted rotor with close-fitting vanes placed in the slots. The rotor is mounted on the motor's driveshaft. It can move within the elliptical cam ring in the motor. Springs or centrifugal forces push the vanes against the cam ring. The motor also has two ports: one serves as the inlet port and the other as the outlet port, depending on the required direction of motion. While the fluid pushes half of the consecutive vanes, the other half helps the motor in discharging the spent fluid through its outlet port. The motor develops the output torque by allowing fluid pressure to act on its vanes.

Vane motors operate well at speeds from 20 to 6000 rpm. Moreover, they are suited for the bidirectional operation. They can operate at significantly lower noise levels than other types of hydraulic motors. The cost of vane motors is higher than that of gear motors but lower than that of piston motors, for comparable power ratings.

However, the vane motors are affected by the higher rate of fluid contamination. They are susceptible to higher internal leakage, especially at low speeds. They are also less efficient than the piston motors. The service life of the vane motors is shorter than that of the piston motors but longer than that of the gear motors.

Vane motors are extensively used on machine tools, plastic molding machines, winches, and hydrostatic transmissions. However, they are rarely used in mobile applications outside high-speed and low-speed drilling.

Piston Motors

Piston motors are also positive-displacement devices. A piston motor consists of a cylinder block with many pistons, a cam/swash plate, and a drive shaft. In general, the designs of the piston motors are very similar to those of the piston pumps. However, the operation of the piston motors is just the reverse of that of the piston pumps. That is, every piston pump produces hydraulic power in response to the rotary mechanical power at its drive shaft. In contrast, every piston motor produces rotary mechanical power at its drive shaft in response to the applied hydraulic power.

Piston motors are classified according to various parameters. According to the arrangement of the cylinder blocks with respect to their drive shafts, they are classified as axial piston motors and radial piston motors. In the axial piston motor, the cylinder blocks are arranged axially, whereas in the radial piston motor, the cylinders are arranged radially. Further, based on motor displacement, piston motors can be classified as fixed-displacement or variable-displacement.

Piston motors have a wide speed range, typically operating from 10 rpm to 5000 rpm while maintaining stable torque output. They deliver the highest torque, speed, and power for medium- and heavy-duty applications. They are, in general, the most efficient and versatile hydraulic motors, but they are the most expensive, as compared to other types of hydraulic motors.

Axial Piston Motors

An axial piston motor uses an axially-mounted piston block to generate mechanical power. When the high-pressure fluid from the high-pressure system flows into the motor, the pistons are forced to move within their chambers. This action causes the motor to generate an output torque. Axial piston motors can be classified into in-line piston motors and bent-axis piston motors.

Further, they can be either fixed-displacement or variable-displacement. The fixed-displacement axial piston motor has a stationary cam plate, whereas the variable-displacement unit has mechanical means to vary the angle of its cam plate. The motor torque and speed depend on the cam plate angle.

In general, the axial piston motors have excellent high-speed and speed-reversal capabilities. Moreover, they can accelerate quickly. They can provide excellent starting torque and a smooth, low-speed, high-torque operation. They also have excellent volumetric efficiencies, especially at lower pressures.

In-line Axial Piston Motor

In-line piston motors are the most commonly used rotary actuators in hydraulic systems. It mainly consists of a cylinder block with many pistons, an angled cam/swash plate, inlet and outlet ports, and a drive shaft. The cylinders, as shown in Figure 8.5, are arranged in a circle, parallel to each other. That is, the motor shaft and the cylinder block are aligned along the same axis.

The swash plate is located at one end of the cylinder block and is acted upon by the cylinder pistons.

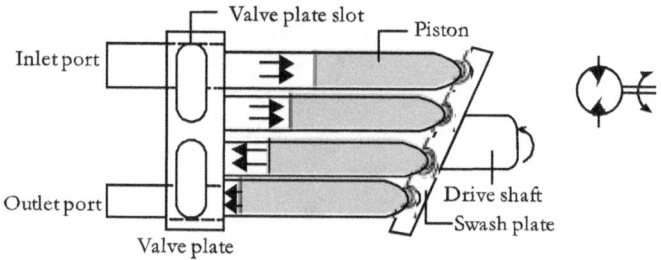

Figure 8.5 | A schematic diagram showing the cross-sectional view of an in-line axial piston motor

The application of high-pressure fluid from the system at the motor inlet port applies pressure to the ends of the cylinder pistons, which then reciprocate within the cylinder block. The cylinders are filled with the fluid in a particular sequence. This action causes the pistons to move outwards, sequentially pushing against the angled swash plate and rotating the cylinder block and the driveshaft. The fluid is then swept back into the system reservoir at low pressure during the piston's return stroke. The torque produced by the in-line axial piston motor is related to the swash plate angle and the area of the pistons.

The in-line axial piston motors are available in fixed-displacement and variable-displacement variants. In the fixed-displacement type, the angle of the swash plate is set. In contrast, in the variable-displacement type, the angle can be varied by various means, ranging from a simple lever to sophisticated servo controls. Increasing the swash plate angle improves the motor's torque capacity but reduces the shaft speed, and vice versa. Most manufacturers recommend a minimum swash plate angle of 15-17° for best results. A maximum angle of 40-45° yields good torque output and extended motor life.

Axial-piston motors are well-known for their high volumetric efficiency in both high- and low-speed operation. They are the most efficient hydraulic motors compared to other types. They work best in the low-speed high-torque (LSHT) applications. In-line axial piston motors find applications in agricultural and construction equipment.

Bent-axis Axial Piston Motor

Figure 8.6 gives the cross-sectional view of a bent-axis piston motor. It consists of a cylinder block with pistons, a cam/swash plate, inlet and outlet ports, and a drive shaft. Unlike the case of the in-line axial piston motor design, the axis of the cylinder block and the axis of the drive shaft in the bent-axis axial piston motor design are arranged at an angle to each other. The torque is developed in the motor as a reaction to the system pressure exerted on its reciprocating pistons. Because the axis of the shaft and the axis of the cylinder block are set at an angle, the force acting on the joint is resolved into axial and tangential components. The bearings take up the axial load while the tangential component develops the torque at the motor's shaft. The bent-axis axial piston motor has the same operating characteristics as the in-line axial piston motor.

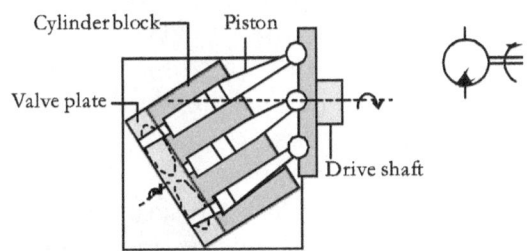

Figure 8.6 | A bent-axis axial piston motor

Bent-axis piston motors are available in fixed-displacement and variable-displacement types. Variable-displacement types can be controlled either mechanically or by pressure compensation. The angle of the cylinder block with the drive shaft of a bent-axis piston motor determines its torque and speed ranges. The greater the angle, the higher the torque and the lower the motor speed.

Bent-axis piston motors are rugged and capable of handling higher operating pressures. Since there is no sliding action of the piston shoes in a bent-axis piston motor, it tends to generate higher torque and lower friction for a specified energy input. However, the bent-axis piston motors are hefty, particularly the variable-displacement type. They find many applications in earthmoving machines, construction equipment, forestry equipment, marine equipment, offshore equipment, industrial conveying systems, heavy-duty winches, and high-power crushers.

Note: Appendix 2 gives the performance data for Bent-axis piston motors.

Radial Piston Motors

Radial piston motors have a wide variety of designs within the primary radial configuration. Figure 8.7 shows the cross-sectional view of one type of radial piston motor for a hydraulic system. It consists of a radially arranged cylinder block with pistons arranged within the housing, a thrust ring, a pintle, and a driveshaft.

The cylinder block with five to seven radial bores is attached to the driveshaft, and the piston in each bore reciprocates when pressurized fluid is applied. The outer piston ends bear against the thrust ring.

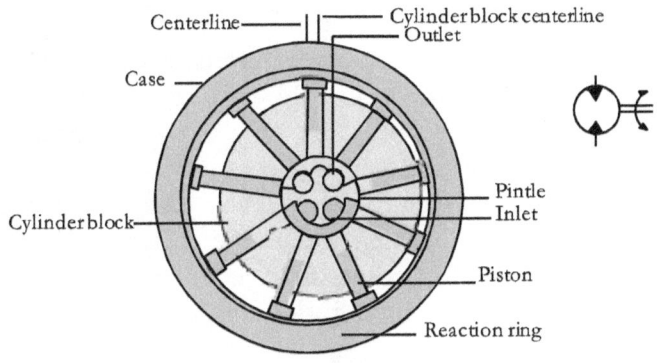

Figure 8.7 | A radial piston motor

The flow of pressurized fluid through the pintle at the motor's central location pushes the pistons outward. As a result, the pistons are pushed against the thrust ring, and the barrel is rotated by the reaction forces developed in the motor.

Motor displacement can be varied by shifting the cylinder block laterally. There is no fluid flow and, therefore, no rotation of the barrel when the centerlines of the cylinder block and the housing coincide. The motor's rotational direction can be reversed by moving the cylinder block from its current position to the opposite side, past the center of the housing.

Radial piston motors are robust, compact, and very efficient. They have excellent low-speed capabilities and long service life. They are well-suited to start under loads. However, they have only limited high-speed capabilities and are costly. LSHT radial piston motors are used in drive applications in compact machines, such as skid steers and mini excavators. They are used as wheel motors and for other suitable applications, such as forklifts.

Mounting of Hydraulic Motors

Side loads on a hydraulic motor can cause excessive wear of its parts. Therefore, the motor should be mounted to avoid side loads on its output shaft. When side loads on the motor are unavoidable, it is necessary to support the motor's output shaft and the attached load with auxiliary bearings. The coupling flange must be correctly aligned to the motor shaft. A flexible shaft coupling can be used whenever possible to avoid side loads caused by shaft misalignment.

Advantages and Disadvantages of Hydraulic Motors

Hydraulic motors have relatively large power-to-weight ratios. That means they assist in developing compact systems. They are also simple and reliable. Each one can provide infinitely variable speed control and stalling capability, even under full-load operation. The motors are also capable of quickly reversing their rotation. The main issues with them in service are seal failures, excessive leakage, and associated noise.

Selection of Hydraulic Motors

The selection of a hydraulic motor that is correctly sized for a given application depends on many factors, including system and motor parameters. For example, speed and torque requirements are two essential parameters to consider when selecting motors. The motor must be sized by considering the system pressure and flow rate. The parameters for the selection of hydraulic motors are the system parameters, such as operating pressure range, flow, fluid viscosity, operating temperature, and noise, and the motor parameters, such as its size, weight, displacement, speed, torque, volumetric and mechanical efficiencies, mounting requirements, cost, and estimated service life.

Chapter 9 | Comparison of Hydraulic Motors

Table 9.1 gives a comparison of hydraulic motors against typical parameters in a generalized way.

Table 9.1: Comparison of hydraulic motors (1/2)

Characteristics	Gear motors	Vane motors	Piston motors
General	Simple and rigid design	Most general-purpose motors	Versatile
Construction	Robust, rugged	More complicated than gear types	With tight tolerance
Size	Very compact	Compact	Heavy, bulky
Pressure	Suited for low-pressure applications	Suited for medium pressures	Suited for high-pressure applications
Speed	500-3000 rpm	100 –4000 rpm	Axial: 10-4500 rpm Radial: 0.1–2000 rpm
Low-speed operation	Not well-suited	Inefficient at low speeds	Excellent choice (Radial piston)

Table 9.1: Comparison of hydraulic motors (2/2)

Characteristics	Gear motors	Vane motors	Piston motors
High-speed operation	Best suited	Right choice	Excellent choice (Axial piston motors)
Duty	Light	Light/Medium	Heavy
Efficiency	Least efficient at low-speed ranges	Not so efficient at low speeds	Most efficient
Cost	Least expensive	Moderate	Most expensive
Slippage	Uniform slippage losses	Higher internal leakage	Provide the best sealing
Noise	Most noisy	Least noisy	Noisy
Contamination tolerance	More tolerant	Less tolerant of contamination	Sensitive to contamination
Service life	Long	Shorter	Long

Chapter 10 | Performance Characteristics of Hydraulic Motors

The performance of a hydraulic motor is influenced by various parameters of the associated system as well as by various motor parameters. The system parameters include maximum pressure (continuous and intermittent), maximum fluid flow, maximum fluid viscosity, and operating temperature. The motor parameters include its displacement, speed (maximum and minimum), operating torque, power, volumetric and mechanical efficiencies, weight, and service life. The performance of a hydraulic motor is also affected by its operating conditions, duty cycle, leakage, and coupling method. The performance can be analyzed using characteristics such as the torque vs. speed curve, the pressure vs. volumetric efficiency curve, and the flow vs. speed curve. The following sections give these characteristics.

Torque-Speed Characteristic

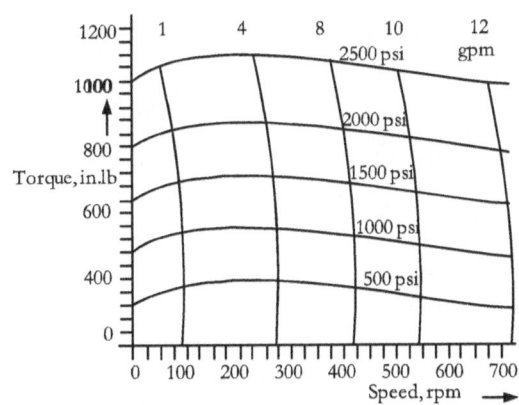

Figure 10.1 | Typical torque-speed characteristics

45

Figure 10.1 shows the typical torque-speed characteristics of a hydraulic piston motor. The motor speed is shown on the x-axis, and its torque on the y-axis. The horizontal curves correspond to the characteristics at different pressures. Sloping vertical curves represent the fluid flow rates through the motor.

Pressure-Volumetric Efficiency Curves

Figure 10.2 shows the pressure–volumetric efficiency curves for the gear, vane, and piston motors for direct comparison. For different types of motors connected to hydraulic systems, fluid leakage increases, and consequently, their efficiencies decrease in varying degrees as the corresponding system pressure increases. The volumetric efficiency of the gear motor decreases linearly with increasing pressure. The volumetric efficiency of the vane motor is typically higher than that of the gear motor. Typically, the vane motor has the highest volumetric efficiency for p < 1000 psi. At higher pressures, the piston motor has higher volumetric and overall efficiencies.

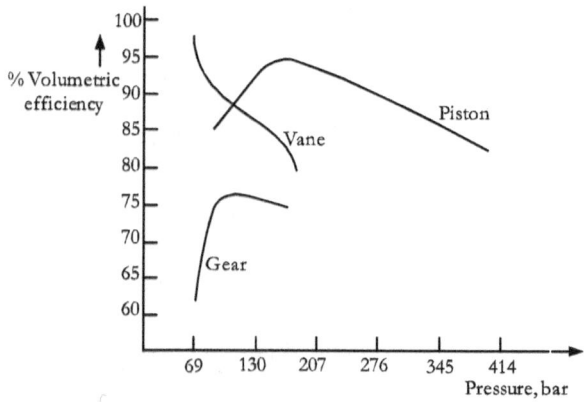

Figure 10.2 | Typical pressure-volumetric efficiency curves

Torque and Flow Curves against Speed

Figure 10.3 shows the typical flow/torque/power characteristics of a hydraulic motor plotted against its speed. The lower line is the flow-versus-speed characteristic, showing a linear relationship between flow rate and speed. It may be noted that the slope of the line is the displacement of the motor. The upper line is the torque Vs speed characteristic. The motor torque tends to drop off at higher speeds, as shown. Also shown in the figure are the motor's output power-versus-speed and flow-versus-speed curves.

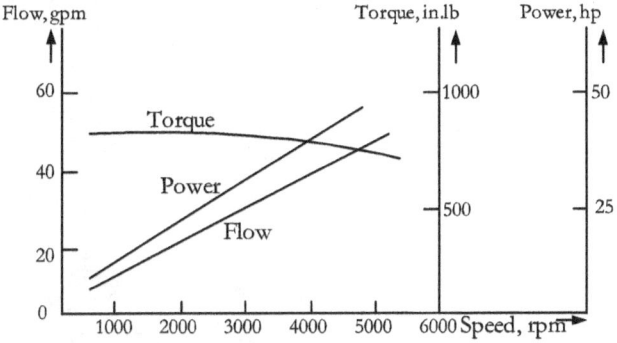

Figure 10.3 | Typical torque-speed and flow vs speed curves

Summary of Performance Characteristics

Hydraulic motors are characterized by high torque at low speeds, seamless speed regulation, and suitability for heavy industrial and mobile applications. They offer a high power-to-weight ratio, making them ideal for applications with limited space. Most hydraulic motors can rotate in both directions, with many capable of instant reversal. Nonetheless, they face limitations, including leakage and sensitivity to temperature changes.

Chapter 11 | Applications of Hydraulic Motors

Hydraulic motors are well-suited to a wide range of applications due to their many advantages. They are widely used for operations such as opening and closing, drilling, mixing, agitating, dumping, feeding, material handling, pushing, pulling, and lowering, as well as for propelling wheeled or track-driven vehicles. They are used in heavy-duty applications, including steel mills, machine tools, agriculture, construction, crane drives, winches, defense, aerospace, marine, and mining. They are also used in military vehicles, excavators, and drilling rigs. Figure 11.1 shows some applications of motors.

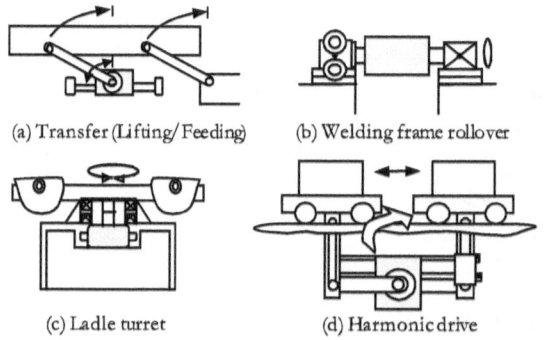

(a) Transfer (Lifting/Feeding)   (b) Welding frame rollover

(c) Ladle turret   (d) Harmonic drive

Figure 11.1 | Typical applications of hydraulic motors

The following texts highlight some industry-specific applications of hydraulic motors. In the steel industry, they tilt electric furnaces. In the construction equipment, they position tools and implements. In material handling applications, they drive conveyors. In aerospace applications, they are used to operate aircraft edge flaps and landing gear. In marine systems, they open and close hatches and operate booms. In the underground mining machinery, they can be used to place a percussion drill.

Chapter 12 | Maintenance of Hydraulic Motors

Most maintenance considerations for hydraulic motors are the same as those for hydraulic pumps. The most common causes of premature hydraulic motor failure are dirt, corrosion, misalignment, loose bolting, and overload. It is also necessary to periodically check the motor's drive shaft for misalignment or damage.

About 70-80% of hydraulic failures are caused by contaminated fluid. Fluid cleanliness can be ensured through visual inspections, servicing, the use of filters, and fluid analysis. Well-maintained filters remove harmful contaminants. It is essential to replace clogged filters regularly. Water content can be reduced using water-removal media or vacuum dehydration. The acid level in the hydraulic fluid should be reduced by controlling the fluid's oxidation rate.

It is essential to control the fluid temperature, as higher temperatures degrade hydraulic fluid and damage seals. Lower temperatures increase fluid viscosity and, hence, lubricity. Fluid deterioration affects the hydraulic motor.

It is also essential to check for leaks around fittings and seals to prevent contamination ingress.

Further, it is necessary to check mounting bolts for tightness and couplings for misalignment.

Measure the case drain flow to detect internal wear and predict potential hydraulic motor failure.

Table 12.1 presents a troubleshooting chart for hydraulic motors.

Table 12.1 | Troubleshooting chart of hydraulic motors

Disturbances	Possible causes	Rectification
Motor rotating in the wrong direction	Incorrect piping between the control valve and the motor	-Connect the circuit as per the circuit diagram
Motor not developing proper speed or torque	Incorrect setting of PRV	-Set PRV correctly
	PRV sticking/open	-Remove dirt under the poppet
	Free circulation of fluid to the reservoir	-Repair or replace the DC valve
	The pump is not delivering sufficient fluid	-Repair pump
Fluid contamination	Filters clogged	-Replace filter elements -Clean strainer -Carry out fluid analysis
External fluid leakage	Worn seals	-Replace worn seals
Fluid is being heated	Inefficiency of the system	-Check the heat exchanger
Misalignment	Mounting loosened	-Tighten the mountings -Re-check the alignment

13 | Objective Type Questions

1. Which of the following statements is <u>incorrect?</u>
a) Gear motors work best in high-speed, low-torque applications.
b) Gear motors have higher volumetric efficiency as compared to other types of motors.
c) Hydraulic vane motors operate at a lower noise level than other types of motors.
d) Piston motors are the most efficient, but most expensive, motors.

2. Which of the following hydraulic motors is more tolerant of contamination?
a) Gear motor
b) Vane motor
c) Axial piston motor
d) Radial piston motor

3. Which of the following hydraulic motors is an excellent choice for high-speed operation?
a) Gear motor
b) Vane motor
c) Axial piston motor
d) Radial piston motor

4. Which power transmission system provides stepless control of speed, torque, and power?
a) Mechanical
b) Electrical
c) Hydrostatic
d) Pneumatic

5. Identify the following component:

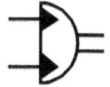

Figure 13.1

a) Semi-rotary pneumatic actuator
b) Semi-rotary hydraulic actuator
c) Hydraulic motor
d) Pneumatic motor

6. Identify the following component:

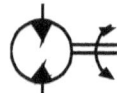

Figure 13.2

a) Semi-rotary pneumatic actuator
b) Semi-rotary hydraulic actuator
c) Hydraulic motor
d) Pneumatic motor

7. Which is the best hydraulic motor for excellent low-speed operation?
a) Gear motor
b) Vane motor
c) axial piston motor
d) Radial piston motor

14 | Review Questions

1. Define hydraulic motors?
2. What are the main reasons for using hydraulic rotary actuators in industrial and mobile hydraulic systems?
3. Explain the working principles of a hydraulic motor.
4. Draw the symbols for the hydraulic semi-rotary actuators and motors.
5. Write two differences between the hydraulic motors and the hydraulic pumps.
6. Define the term 'operating pressure' of a hydraulic motor.
7. What determines the selection of a hydraulic motor operating speed?
8. Define hydraulic motor torque.
9. Define the following terms of hydraulic motors: (a) starting torque, (b) running torque, and (c) stalling torque.
10. Write the formula for calculating the torque of a hydraulic motor.
11. What determines the operating speed of a hydraulic motor?
12. Define the fluid power and the shaft power of a hydraulic motor.
13. Write the formula for calculating the input horsepower in the English units?
14. Write the formula to calculate the mechanical horsepower output of a hydraulic motor in the English units?
15. Define the nominal displacement of hydraulic motors.
16. Explain the relationship between the flow rate and the shaft speed of a hydraulic motor.
17. What is an ideal hydraulic motor?
18. What factors determine the torque output of a hydraulic motor?

19. Define the following terms of hydraulic motors: (1) Motor displacement, (2) Volumetric efficiency, and (3) Mechanical efficiency.

20. Define and explain the terms concerning hydraulic motors: (1) Volumetric efficiency, and (2) overall efficiency.

21. Write the formula for calculating the overall efficiency of a hydraulic motor.

22. Explain the interaction of the flow rate and the pressure during the operation of a hydraulic motor.

23. Determine the power, torque, and flow rates of hydraulic motors.

24. What are the functions of hydraulic rotary seals?

25. Give a brief note on the case drain connection in hydraulic motors.

26. List two differences between the hydraulic pumps and the hydraulic motors

27. How are hydraulic motors classified?

28. What are the different ways of constructing hydraulic motors?

29. How can the output speed of a variable-displacement motor be changed?

30. Name two types of positive-displacement hydraulic motors.

31. Describe the general constructional features of one type of semi-rotary hydraulic actuator.

32. Explain the working of a vane type semi-rotary hydraulic actuator with a simple sketch.

33. Explain the working of a rack-and-pinion type semi-rotary hydraulic actuator with a sketch.

34. Explain the operation of a gear motor.

35. Describe the general constructional features of gear motors.

36. Give two advantages and disadvantages of gear motors.

37. What are the applications of gear motors?

38. Explain the operation of a gerotor motor.

39. Describe the general constructional features of gerotor motors.

40. What are the advantages and disadvantages of gerotor motors?

41. What are the applications of gerotor motors?
42. Explain the construction features of a Geroler motor.
43. What are the typical characteristics of Geroler motors?
44. Describe the operation of a hydraulic vane motor.
45. Describe the general constructional features of hydraulic vane motors.
46. Give the advantages and disadvantages of hydraulic vane motors?
47. What are the applications of hydraulic vane motors?
48. Explain the difference between the vane motor and the vane pump, as used in a hydraulic system.
49. Explain the working of an in-line axial piston motor.
50. Describe the general constructional features of in-line axial piston motors.
51. What are the advantages and disadvantages of in-line axial piston motors?
52. What are the applications of in-line motors?
53. How is the fluid displacement adjusted in a variable-displacement axial-piston motor?
54. Explain the operation of a bent-axis axial piston motor.
55. Describe the general constructional features of bent-axis axial piston motors.
56. What are the advantages and disadvantages of bent-axis axial piston motors?
57. What are the applications of bent-axis axial piston motors?
58. Explain the operation of a radial piston motor.
59. Describe the general constructional features of radial piston motors.
60. What are the advantages of radial piston motors?
61. What are the applications of radial piston motors?
62. Draw the torque-speed characteristics of a hydraulic motor and explain.
63. Draw the pressure-volumetric efficiency curves of the gear, vane, and piston motors and compare.
64. Write two factors that affect the rating and selection of a hydraulic motor for an application.
65. Briefly explain the areas of application of hydraulic motors.

15 | Numerical Problems

1. A hydraulic motor theoretically consumes 48 in³/min while running at a speed of 2000 rpm. What is the volumetric displacement of the motor? [Ans: 0.024 in³/rev]

2. A hydraulic piston motor operates with a pressure drop of 4000 psi across its ports. Measured flow rate to the motor is 52.8 gpm. What is the input hydraulic power? [Ans: 123 hp]

3. A hydraulic piston motor must produce an output power of 104.56 hp with a pressure differential of 2900 psi. The overall efficiency is 91%. What is the flow rate of fluid required? [Ans: 67.9 gpm]

4. What is the output power of a hydraulic vane motor if the measured torque produced by the motor is 531 in.lb while running at 1000 rpm? [Ans: 8.43 hp]

5. Calculate the output power of the hydraulic gear motor that has an actual flow rate of 79.3 gpm and a pressure differential of 1160 psi. The overall efficiency is 65%. [Ans: 34.88 hp]

6. What is the theoretical output speed of a hydraulic motor with a displacement of 4 in³/rev and a steady flow rate of 10 gpm? If the actual speed measured is 500 rpm, what is the motor's volumetric efficiency? [Ans: 6.5%]

7. A hydraulic motor operates with a pressure drop of 2000 psi across its ports. The measured flow rate to the motor is 10 gpm. What is the input hydraulic power? If the measured torque is 1080 in.lb at 536 rpm, what is the output power? Also, calculate the overall efficiency.
[Ans: P_{in}= 11.67 hp, P_{out} =9.19 hp, η_o= 78%]

8. The torque required for turning a sugar mill drive using a hydraulic motor is 1814 in.lb. The drive unit is powered by a power pack delivering 55.5 gpm at 2990 psi. Assume a mechanical efficiency of the motor as 88%. What is the required displacement and speed of the hydraulic motor?
[Ans: V_D=4.46 in³/rev, N=2870 rpm]

Objective-type questions - answer key:
1-b, 2-a, 3-c, 4-c, 5-b, 6-c, 7-d

Appendix 1

The data is extracted from the actual catalogue listing of some prominent manufacturers.

Performance Data of Geroler Motors for continuous duty

Table A1.1 | Values of Torque in lb.in (Speed in rpm)

ΔP(psi) Q(gpm)	$V_D = 0.50$ in³/rev						
	200	400	600	800	1000	1500	2030
1	11 (456)	25 (444)	40 (429)	55 (412)	69 (394)	102 (332)	132 (239)
2	9 (897)	24 (886)	38 (867)	53 (847)	68 (823)	105 (749)	141 (647)
3	6 (1349)	20 (1331)	35 (1309)	51 (1285)	65 (1261)	102 (1176)	139 (1060)
4.25		16 (1902)	30 (1873)	44 (1846)	60 (1817)	97 (1721)	135 (1585)
4.5		16 (1992)	29 (1964)	43 (1929)	59 (1900)	96 (1808)	134 (1673)

Table A1.2 | Values of Torque in lb.in (Speed in rpm)

ΔP(psi) Q(gpm)	$V_D = 0.79$ in³/rev						
	200	400	600	800	1000	1500	2030
1	19 (290)	43 (285)	65 (277)	88 (268)	109 (260)	164 (230)	217 (189)
2	16 (573)	39 (566)	63 (555)	86 (544)	109 (534)	165 (490)	225 (437)
3	11 (859)	35 (849)	58 (838)	82 (825)	105 (810)	163 (763)	223 (701)
4	6 (1153)	30 (1140)	53 (1129)	76 (1117)	99 (1101)	157 (1044)	217 (975)
5.5		19 (1575)	42 (1556)	65 (1539)	89 (1521)	148 (1457)	209 (1387)

Table A1.3 | Values of Torque in lb.in (Speed in rpm)

ΔP(psi) Q(gpm)	$V_D = 1.21$ in³/rev						
	200	400	600	800	1000	1500	2030
1	32 (189)	67 (187)	102 (185)	136 (182)	170 (179)	253 (169)	325 (138)
2	30 (379)	65 (375)	101 (370)	136 (366)	172 (361)	257 (347)	333 (309)
3	21 (569)	57 (565)	93 (560)	128 (556)	163 (551)	248 (523)	330 (484)
4	12 (761)	47 (758)	83 (751)	119 (746)	154 (741)	239 (707)	320 (656)
5.5		31 (1043)	67 (1035)	101 (1028)	137 (1021)	218 (990)	299 (934)

Table A1.4 | Values of Torque in lb.in (Speed in rpm)

ΔP(psi) Q(gpm)	$V_D = 1.93$ in³/rev						
	200	400	600	800	1000	1500	1750
1	51 (118)	106 (116)	160 (113)	213 (111)	265 (107)	383 (81)	439 (70)
2	46 (236)	103 (234)	159 (230)	214 (225)	269 (221)	387 (175)	446 (165)
3	36 (355)	94 (352)	149 (347)	205 (342)	259 (336)	377 (287)	440 (273)
4	24 (474)	79 (472)	135 (466)	190 (460)	246 (452)	362 (393)	425 (373)
5.5		55 (650)	111 (645)	167 (636)	221 (629)	334 (575)	400 (550)

Table A1.5 | Values of Torque in lb.in (Speed in rpm)

ΔP(psi) Q(gpm)	$V_D = 3.0$ in³/rev						
	200	400	600	800	1000	1200	1400
1	82 (75)	167 (72)					
2	70 (149)	156 (147)	243 (144)	327 (142)			
3	53 (221)	140 (220)	227 (217)	311 (213)	396 (209)	484 (201)	549 (191)
4	30 (296)	120 (292)	204 (286)	292 (282)	374 (273)	460 (265)	541 (259)
5.5		81 (393)	170 (389)	254 (383)	339 (377)	422 (369)	506 (358)

Table A1.6 | Values of Torque in lb.in (Speed in rpm)

$V_D = 2.2$ in³/rev							
ΔP(psi) Q(gpm)	200	400	600	800	1000	1400	1800
2	49 (204)	103 (201)	162 (198)	189 (194)	270 (189)	379 (177)	489 (162)
4	47 (408)	106 (407)	160 (402)	191 (399)	274 (394)	384 (381)	495 (365)
6	44 (613)	102 (612)	158 (609)	188 (604)	272 (599)	383 (586)	496 (565)
8	40 (817)	97 (817)	153 (814)	184 (807)	270 (799)	383 (785)	497 (762)
10	36 (1021)	90 (1021)	148 (1015)	180 (1008)	265 (1001)	380 (981)	495 (959)

Table A1.7 | Values of Torque in lb.in (Speed in rpm)

$V_D = 2.8$ in³/rev							
ΔP(psi) Q(gpm)	200	400	600	800	1000	1400	1800
2	64 (161)	136 (158)	212 (156)	284 (153)	355 (148)	497 (139)	641 (127)
4	61 (323)	139 (320)	209 (316)	286 (314)	359 (310)	503 (300)	649 (287)
6	58 (486)	134 (481)	207 (479)	282 (475)	356 (471)	502 (461)	650 (444)
8	52 (648)	128 (643)	200 (640)	276 (635)	354 (628)	502 (617)	651 (599)
10	47 (808)	118 (803)	194 (798)	269 (793)	347 (787)	498 (771)	649 (753)
12	36 (969)	109 (964)	188 (960)	260 (952)	340 (946)	492 (931)	643 (914)

Table A1.8 | Values of Torque in lb.in (Speed in rpm)

$\Delta P(psi)$ Q(gpm)	$V_D = 3.6$ in³/rev						
	200	400	600	800	1000	1400	1800
2	79 (127)	169 (125)	260 (123)	305 (121)	437 (117)	616 (109)	796 (96)
4	76 (254)	168 (254)	257 (251)	307 (249)	441 (246)	620 (236)	800 (224)
6	73 (381)	161 (381)	252 (380)	303 (377)	439 (373)	618 (364)	802 (349)
8	64 (508)	151 (508)	243 (508)	294 (504)	428 (500)	609 (491)	794 (476)
10	57 (635)	141 (635)	234 (634)	283 (630)	419 (626)	602 (614)	786 (601)
12	45 (762)	131 (762)	227 (762)	274 (757)	409 (753)	593 (741)	778 (728)
14	33 (889)	118 (889)	213 (887)	266 (882)	396 (877)	583 (866)	770 (851)
15	29 (953)	111 (953)	205 (951)	260 (945)	389 (940)	576 (929)	765 (913)

Table A1.9 | Values of Torque in lb.in (Speed in rpm)

$\Delta P(psi)$ Q(gpm)	$V_D = 4.5$ in³/rev						
	200	400	600	800	1000	1400	1800
2	103 (101)	220 (99)	339 (98)	454 (96)	569 (93)	801 (86)	1036 (76)
4	99 (203)	219 (201)	335 (199)	457 (197)	574 (194)	808 (187)	1042 (177)
6	94 (305)	210 (303)	328 (301)	451 (298)	571 (296)	805 (288)	1044 (276)
8	86 (406)	196 (404)	319 (402)	438 (399)	558 (396)	793 (388)	1033 (377)
10	74 (507)	183 (505)	310 (502)	422 (499)	545 (496)	784 (486)	1024 (476)
12	58 (608)	171 (606)	295 (603)	408 (600)	533 (596)	773 (587)	1013 (576)
14	43 (709)	154 (706)	277 (702)	396 (698)	515 (694)	760 (686)	1002 (674)
15	36 (760)	145 (757)	268 (753)	387 (749)	506 (744)	750 (735)	996 (723)

Table A1.10 | Values of Torque in lb.in (Speed in rpm)

ΔP(psi) Q(gpm)	$V_D = 5.9$ in³/rev						
	200	400	600	800	1000	1400	1800
2	134 (78)	292 (76)	442 (75)	593 (73)	746 (71)	1054 (65)	1365 (55)
4	131 (156)	281 (155)	436 (153)	596 (151)	750 (149)	1059 (143)	1367 (134)
6	126 (234)	269 (233)	425 (231)	588 (230)	747 (228)	1054 (221)	1368 (210)
8	110 (312)	246 (311)	408 (310)	566 (308)	718 (305)	1023 (300)	1339 (291)
10	96 (390)	231 (389)	392 (387)	539 (385)	699 (383)	1005 (376)	1318 (368)
12	77 (468)	218 (467)	378 (465)	522 (463)	681 (460)	990 (453)	1301 (445)
14	60 (546)	197 (544)	358 (542)	513 (539)	662 (537)	973 (531)	1293 (521)
15	52 (585)	189 (583)	346 (581)	495 (578)	651 (575)	963 (589)	1286 (559)

Table A1.11 | Values of Torque in lb.in (Speed in rpm)

ΔP(psi) Q(gpm)	$V_D = 7.3$ in³/rev						
	200	400	600	800	1000	1400	1800
2	162 (62)	357 (61)	544 (61)	736 (59)	927 (58)	1305 (53)	1687 (45)
4	160 (125)	348 (124)	539 (123)	736 (121)	930 (120)	1316 (116)	1698 (110)
6	155 (188)	338 (187)	530 (186)	729 (185)	923 (183)	1310 (178)	1699 (170)
8	139 (250)	319 (250)	515 (249)	710 (247)	901 (245)	1283 (241)	1673 (233)
10	121 (313)	303 (312)	497 (311)	686 (309)	883 (308)	1267 (302)	1655 (296)
12	102 (375)	288 (374)	480 (373)	664 (371)	862 (370)	1246 (365)	1640 (358)
14	78 (438)	263 (437)	458 (435)	652 (433)	841 (431)	1228 (427)	1616 (419)
15	67 (469)	253 (468)	446 (466)	632 (464)	828 (462)	1214 (458)	1608 (450)

Table A1.12 | Values of Torque in lb.in (Speed in rpm)

ΔP(psi) Q(gpm)	$V_D = 8.9 \ in^3/rev$						
	200	400	600	800	1000	1400	1700
2	198 (51)	435 (50)	664 (50)	897 (49)	1130 (47)	1591 (43)	1942 (39)
4	196 (103)	424 (102)	657 (101)	898 (99)	1133 (99)	1604 (95)	1954 (92)
6	189 (154)	412 (153)	646 (152)	889 (151)	1125 (150)	1598 (146)	1951 (141)
8	169 (205)	389 (205)	628 (204)	866 (203)	1098 (201)	1564 (197)	1919 (193)
10	148 (257)	369 (256)	605 (255)	836 (253)	1076 (252)	1544 (248)	1899 (244)
12	125 (308)	351 (307)	586 (306)	810 (305)	1051 (303)	1519 (299)	1878 (295)
14	95 (359)	321 (358)	558 (357)	795 (355)	1026 (354)	1497 (350)	1851 (346)
15	82 (385)	308 (384)	544 (383)	771 (381)	1010 (379)	1480 (375)	1840 (371)

Table A1.13 | Values of Torque in lb.in (Speed in rpm)

ΔP(psi) Q(gpm)	$V_D = 9.7 \ in^3/rev$						
	200	400	600	800	1000	1400	1650
2	209 (47)	465 (46)	715 (46)	973 (45)	1228 (44)	1724 (40)	2046 (37)
4	210 (94)	460 (94)	710 (93)	971 (91)	1229 (91)	1745 (89)	2059 (87)
6	205 (141)	454 (141)	704 (140)	965 (139)	1216 (138)	1738 (134)	2055 (132)
8	186 (188)	440 (188)	693 (187)	951 (186)	1205 (185)	1716 (181)	2038 (178)
10	164 (235)	422 (234)	671 (234)	930 (232)	1189 (232)	1702 (228)	2032 (225)
12	144 (282)	404 (281)	652 (281)	900 (279)	1163 (279)	1674 (275)	2004 (272)
14	109 (330)	374 (329)	623 (328)	883 (327)	1140 (325)	1653 (322)	1963 (319)
15	92 (353)	359 (352)	612 (351)	861 (350)	1123 (348)	1633 (345)	1950 (342)

Table A1.14 | Values of Torque in lb.in (Speed in rpm)

ΔP(psi) Q(gpm)	$V_D = 11.3\ in^3/rev$						
	200	400	600	800	1000	1400	1700
2	257 (40)	554 (40)	847 (39)	1150 (38)	1447 (37)	2035 (33)	2320 (29)
4	254 (81)	546 (81)	845 (80)	1145 (79)	1448 (78)	2049 (76)	2343 (74)
6	246 (121)	540 (121)	834 (120)	1137 (120)	1434 (119)	2036 (115)	2337 (112)
8	224 (162)	520 (162)	820 (161)	1117 (160)	1414 (159)	2014 (155)	2315 (152)
10	202 (202)	499 (202)	793 (201)	1095 (201)	1394 (200)	1997 (196)	2299 (193)
12	176 (243)	475 (242)	767 (242)	1063 (241)	1368 (240)	1969 (236)	2268 (234)
14	140 (283)	443 (283)	735 (282)	1035 (281)	1340 (280)	1936 (277)	2227 (274)
15	120 (304)	425 (303)	719 (302)	1014 (301)	1320 (300)	1914 (297)	2205 (294)

Table A1.15 | Values of Torque in lb.in (Speed in rpm)

ΔP(psi) Q(gpm)	$V_D = 14.1\ in^3/rev$						
	200	400	600	800	1000	1400	1450
2	338 (32)	707 (32)	1074 (31)	1456 (30)	1827 (30)	2572 (26)	2657 (25)
4	328 (65)	695 (65)	1076 (64)	1447 (63)	1827 (62)	2577 (60)	2669 (60)
6	317 (97)	687 (97)	1057 (97)	1434 (96)	1811 (95)	2555 (92)	2650 (91)
8	289 (130)	659 (130)	1038 (130)	1406 (129)	1777 (128)	2531 (124)	2625 (124)
10	265 (162)	631 (162)	1004 (162)	1381 (162)	1751 (160)	2510 (156)	2602 (156)
12	230 (195)	599 (195)	968 (194)	1345 (194)	1722 (193)	2480 (189)	2571 (189)
14	191 (227)	563 (227)	927 (227)	1299 (226)	1686 (226)	2428 (222)	2519 (221)
15	167 (243)	538 (243)	904 (243)	1279 (242)	1661 (242)	2404 (238)	2493 (238)

Table A1.16 | Values of Torque in lb.in (Speed in rpm)

ΔP(psi) Q(gpm)	$V_D = 17.9$ in³/rev						
	200	400	600	800	1000	1200	1350
2	427 (26)	893 (25)	1361 (25)	1829 (24)	2293 (22)	2672 (16)	2977 (13)
4	419 (51)	886 (51)	1362 (51)	1833 (50)	2305 (49)	2771 (47)	3110 (44)
6	402 (77)	872 (77)	1342 (76)	1819 (76)	2291 (74)	2757 (71)	3098 (68)
8	367 (102)	838 (102)	1316 (102)	1785 (101)	2252 (100)	2723 (98)	3070 (95)
10	332 (128)	803 (128)	1276 (128)	1749 (127)	2215 (126)	2684 (123)	3034 (120)
12	289 (153)	760 (153)	1230 (153)	1706 (153)	2177 (151)	2634 (149)	2989 (146)
14	241 (179)	712 (179)	1176 (179)	1650 (179)	2126 (177)	2592 (175)	2935 (172)
15	211 (192)	683 (192)	1149 (192)	1623 (191)	2096 (190)	2558 (188)	2905 (185)

Table A1.17 | Values of Torque in lb.in (Speed in rpm)

ΔP(psi) Q(gpm)	$V_D = 22.6$ in³/rev						
	200	400	600	800	1000	1200	1250
2	537 (20)	1121 (20)	1715 (20)	2285 (19)	2862 (16)		
4	532 (40)	1123 (40)	1715 (40)	2308 (39)	2893 (38)	3467 (36)	3604 (35)
6	508 (61)	1100 (61)	1693 (61)	2294 (60)	2884 (58)	3458 (55)	3598 (53)
8	463 (81)	1060 (81)	1661 (81)	2255 (80)	2840 (79)	3414 (76)	3557 (74)
10	414 (101)	1017 (101)	1613 (101)	2203 (101)	2788 (99)	3363 (96)	3506 (94)
12	363 (121)	960 (121)	1553 (121)	2152 (121)	2737 (119)	3305 (116)	3446 (115)
14	303 (142)	897 (142)	1484 (142)	2086 (142)	2667 (140)	3246 (137)	3386 (136)
15	266 (152)	862 (152)	1452 (152)	2050 (152)	2630 (150)	3206 (148)	3347 (147)

Table A1.18 | Values of Torque in lb.in (Speed in rpm)

$V_D = 45.1$ in^3/rev			
ΔP(psi) Q(gpm)	200	400	600
2	1080 (10)	2250 (10)	3440 (10)
4	1070 (20)	2250 (20)	3440 (19)
6	1020 (30)	2200 (30)	3390 (29)
8	945 (40)	2135 (40)	3330 (39)
10	840 (50)	2050 (50)	3250 (48)
12	740 (60)	1945 (59)	3130 (58)
14	630 (69)	1820 (68)	3005 (68)
15	540 (74)	1735 (74)	2905 (73)

Appendix 2

Typical Specifications – Bent-axis Axial Piston Motors

Table A2.1 |

Displacement, in³/rev	Maximum pressure, psi (Continuous)	Maximum speed, rpm	Motor continuous input flow, gpm
0.299	5000	10800	10.83
0.598	5000	9900	17.70
0.873	5000	9000	22.99
1.159	5000	8100	27.21
1.830	5000	5600	44.39
2.440	5000	5000	52.84
3.649	5000	4300	67.90
4.906	5000	4000	85.07
6.719	5000	3600	104.62
9.154	5000	2600	103.04
14.768	5000	2400	153.24

16 | References

1. Andrew Parr, Hydraulics & Pneumatics, A technician's and engineer's guide, 2nd Edition, Butterworth, Heinemann, 1998

2. Anthony Esposito, Fluid Power with Applications, 6th Edition, Prentice-Hall of India, 2006

3. Article on 'About Hydraulic Motors', GlobalSpec Inc., 350 Jordan Rd, Troy, NY, USA

4. Article on 'Bigger Isn't Always Better When It Comes to Sizing Your Hydraulic Motors for Efficiency', By Phillip Groves, White Hydraulics Inc., Hopkinsville, Ky., in Compact Equipment, October 2003

5. Article on 'How Does a Hydraulic Motor Work', by Shashank Nakate in Buzzle.com

6. Article on 'Hydraulic motors – Part 1' and 'Hydraulic motors – Part 2', Penton Media, Inc., & Hydraulics & Pneumatics magazine

7. Catalogue on Low Speed, High Torque Motors (Document No. E-MOLO-MC001-E9) EATON Corporation, USA

8. Eaton Hydraulics Training Services, Mobile Hydraulic Manual, 2010

9. Publications Department of Womac Machine Supply Company, Industrial Fluid Power, Volume 3, Third Edition

Fluid Power Educational Series Books

1. Pneumatic Systems and Circuits -Basic Level (In the SI Units)
2. Industrial Pneumatics -Basic Level (In the English Units)
3. Pneumatic Systems and Circuits -Advanced Level
4. Electro-Pneumatics and Automation
5. Design of Pneumatic Systems (In the SI Units)
6. Design Concepts in Pneumatic Systems (In the English Units)
7. Maintenance, Troubleshooting, and Safety in Pneumatic Systems
8. Industrial Hydraulic Systems and Circuits -Basic Level (In the SI Units)
9. Industrial Hydraulics -Basic Level (In the English Units)
10. Hydraulic Fluids
11. Hydraulic Filters: Construction, Installation Locations, and Specifications
12. Hydraulic Power Packs (In the SI Units)
13. Power Packs in Hydraulic Systems (In the English Units)
14. Hydraulic Cylinders (In the SI Units)
15. Hydraulic Linear Actuators (In the English Units)
16. Hydraulic Motors (In the SI Units)
17. Hydraulic Rotary Actuators (In the English Units)
18. Hydraulic Accumulators and Circuits (In the SI Units)
19. Accumulators in Hydraulic Systems (In the English Units)
20. Hydraulic Pipes, Tubes, and Hoses (In the SI Units)
21. Pipes, Tubes, and Hoses in Hydraulic Systems (In the English Units)
22. Design of Industrial Hydraulic Systems (In the SI Units)
23. Design Concepts in Industrial Hydraulic Systems (In the English Units)

24. Maintenance, Troubleshooting, and Safety in Hydraulic Systems
25. Hydrostatic Transmissions (HSTs) (In the SI Units)
26. Concepts of Hydrostatic Transmissions (In the English Units)
27. Load Sensing Hydraulic Systems (In the SI Units)
28. Concepts of Load Sensing Hydraulic Systems (In the English Units)
29. Electro-hydraulic Proportional Valves
30. Electro-hydraulic Servo Valves
31. Cartridge Valves
32. Electro-hydraulic Systems and Relay Circuits
33. Practical Book: Pneumatics - Basic Level
34. Practical Book: Electro-pneumatics - Basic Level
35. Practical Book: Industrial Hydraulics – Basic Level
36. Programmable Logic Controllers and Programming Concepts
37. Compressed Air Dryers
38. Hydraulic Circuits – Identification of Components and Analysis

For more details, please visit: **https://jojibooks.com.**

About the Author

Joji Parambath has been a trainer in Pneumatics, Hydraulics, and PLCs for over 25 years. During his career, he has trained numerous industry professionals, as well as faculty members and students from engineering institutions.

At present, he is the key trainer at Fluidsys Training Centre, Bangalore, India (https://fluidsys.org), which provides training in Pneumatics and Hydraulics. He has already written two books on Pneumatics and Hydraulics. The publication of the present series of 32 books is intended to restructure and update the existing books.

The author wishes to thank all trainees for their lively interaction and many useful suggestions during the training programs that prompted the author to write the present series of books. You may send your feedback to joji.p@hotmail.com

10th June, 2020

Further Information
The author has also written six more books since the first edition, bringing the total to thirty-eight published via Amazon Kindle Direct Publishing (KDP).

28th February, 2026